TRADITIONAL, HERBAL MEDICINE

A Global Look at Traditional, Herbal Medicine, Natural Remedies, and Age-Old Healing.

by

J. Radeino Ambe

Sahel Zone LLC

Acknowledgements

To my children, who are now young adults;
***Abba**, **Nabila** and **Mimi**.*

You all are precious gifts in your unique ways. Thank you for your patience as I am now balancing work and my myriad of volunteer projects.

You have encouraged me to get out of my mundane comfort zone and reach for more. For that I am eternally grateful.

A Global Look at Traditional Herbal Medicine, Natural Remedies, and Age-Old Healing

Table of Contents

Background

Many people across the world use traditional medicine, but what is *traditional medicine*?

Traditional medicine may be used for various reasons and in a myriad of settings. Some people choose to use traditional and/or plant-based herbal medicines because it has time-proven evidence and is preferred for treatments. In some countries, it is used out of necessity due to multiple reasons. For instance, people who use it may lack access to a modern medicine health care facility in their city or town or in some cases, people may simply not have the finances to purchase other types of "modern" or "conventional" medicine. The number of people seeking out traditional medicine has increased in recent years as people look for natural remedies that do not have the negative side effects that are exhibited by many synthetic drugs. Despite this, however, modern medicine is looked upon as the Gold Standard, as it were, for treating illnesses and diseases by people in some regions who may not be aware of and/or are yet to gain an understanding of the source of where these medicines were derived or originated.

This book aims to explore the many ways traditional medicines across the world have been used throughout history and continue to be used today. The United States National Cancer Institute defines herbal medicine as *"A type of medicine that uses roots, stems, leaves, flowers, or seeds of plants to improve health, prevent disease, and treat illness"* (National Cancer Institute, n.d.)[1]. For the purposes of this book, the terms "plant-based" and "herbal" may be used interchangeably.

---

[1] National Cancer Institute. (n.d.), Retrieved from https://www.sciencedirect.com/science/article/abs/pii/B978032390572500007X

Although the elements of nature, water, earth, air or wind, and fire are part of traditional medicine, they are not mentioned in depth in the book. The concept of the elements is cross-cutting for obvious reasons. From land comes metals and mud which is often used in rituals. Also in land are minerals, some of which are added to herbal medicine. Springs and rivers frequently had divine spirits who guided the fishermen. Rain and water have been used for cold baths, steam baths, hydrotherapy and the like.

Introduction

According to the World Health Organization definition, traditional medicine is defined as,

.. the sum total of the knowledge, skill, and practices based on the theories, beliefs, and experiences indigenous to different cultures, whether explicable or not, used in the maintenance of health as well as in the prevention, diagnosis, improvement or treatment of physical and mental illness.

WHO Traditional Medicine Strategy 2014-2023.[2]

A trip across the globe from east to west shows how prevalent traditional herbal medicines have been across humankind to heal the sick. Evidence from caves shows that since prehistoric times, traditional medicines were used when people lived in caves, to combat diseases and ailments. Evidence found, such as those associated with burial chambers, have been corroborated by anthropologists and through this, we are able to decipher the role and practice of the resident healer or shaman.

Creating a visual journey from the east in Oceana, where the Indigenous Peoples in New Zealand and Australia are well known for their plant-based medicinal cures, one can see the influence of traditional healing practices that still exist. Moving further across the globe, with a look at cultures in Asia, Chinese traditional medicine dominates. It has influences that are far-reaching across Asia to Japan with *Kampo* medicine, as it is termed in the region. Also, other forms of Chinese medical practices, such as acupuncture, herbology and food therapy, come from the ancient Chinese traditional practitioners.

---

[2] World Health Organization, (2013). Traditional Medicine Strategy 2014-2023. Retrieved from https://iris.who.int/bitstream/handle/10665/92455/9789241506090_eng.pdf?sequence=1

If we follow the trajectory from South East Asia to South Asia, one will invariably find Ayurveda, which is primarily practiced in India and Nepal.

Across Europe, where the Celtic communities dominated central and western Europe as far back as Neolithic times, scientific evidence of the herbs used for medicine has been found. This has also been recorded in Greco-Roman texts where the herbs used for medicine have been documented. Egyptians had extensive knowledge of science and medical practices. Medical papyri examined show the use of different substances, plant and animal to treat illnesses in other regions of Africa. Traditional medicines and herbal remedies have been used often in connection with spells and incantations. In the Americas, the Indigenous People from the Americans (including South America and the Amazon basin) have deep knowledge of the plants in the region. They have been practicing traditional medicine passed down through generations of shamans for centuries.

Chapter 1: Historical Perspectives of Traditional Medicine

From ancient Chinese and Ayurvedic texts to Greek sources, the history of traditional medicine is brimming with interesting information. From Hippocrates to modern-day doctors, this book will explore how these medicines have been used throughout the ages, from their theories to their application in various treatments. With such deep roots in various cultures, it's easy to see why traditional medicine continues to be a popular option when it comes to treating illness and promoting health.

This chapter will move from east to west to describe in more detail the forms of traditional medicine that have evolved over the years, starting with Oriental Medicine. Traditional medicine is a holistic approach to healing that has been used for centuries, dating back as far as ancient Chinese and Ayurvedic texts. This type of medicine uses natural methods such as herbal remedies and acupuncture to treat illnesses, promote wellness, and improve overall health and well-being. Although it has its roots in older cultures, traditional healing practices are still very much alive today due to their proven efficacy in promoting optimal physical, mental, emotional, and spiritual health.

The history of traditional medicine stretches back to the earliest recorded civilizations. In ancient times, people believed that the gods had bestowed upon them specific medicinal knowledge, which was passed down through generations by oral traditions and cultural practices. This knowledge included herbal remedies for treating illness as well as methods for maintaining health. For example, ancient Egyptian medical texts from around 3000 BC already describe the use of various plants and herbs for managing ailments such as headaches, digestive issues, skin conditions, and more.

Traditional Chinese Medicine

Oriental medicine, also known as Traditional Chinese Medicine (TCM), is a system of medicine that is used to treat the "whole person" (body, mind and spirit). It is rooted in our life force or energy known as Qi, pronounced Chi (氣). Chi flows through the body along certain meridians that have been charted. When Yin and Yang are out of balance, this can block the flow of Qi and lead to disease. Traditional Chinese Medicine has its roots in ancient China over two thousand years ago. Historians believe that the foundations of TCM were developed during this time, though the practices have been continuously refined and developed since then. In China, traditional medicine has a long history. The earliest known text on Chinese medicine is the Huangdi Neijing (Yellow Emperor's Inner Canon). This text outlines the theories behind preventative care, diagnosis, and treatment of illnesses using herbal remedies and acupuncture - an integral part of traditional Chinese medicine that is still practiced today.

The famous healer Hua Tuo used acupuncture to treat a variety of conditions alongside herbal remedies and exercises like Tai Chi. Later, the renowned physician Sun Simiao wrote extensively on TCM and added a few innovations of his own.

The Chinese Emperor's court also played an important role in the development of TCM. Emperors employed physicians to treat their families, courtiers, and soldiers. Over time, many of these physicians developed new theories and treatments, which they put into action in the court. By the 19th century, TCM had become a well-established medical system with its own theories and treatments. Despite its long history and prevalence in Chinese culture, TCM was not widely accepted outside of China for many centuries. It began to gain more recognition during the 20th century when practitioners started traveling abroad to teach it to foreign students.

Kampo Medicine in Japan

Kampo is a unique, traditional Japanese medicine that is rooted in ancient Chinese traditional medicinal practices. It evolved from the roots of ancient Chinese medicine and developed further through the climate and culture of the Japanese people into distinctly Japanese traditional medicine. It went through rigorous development in the 17th Century. It is an organized medical system based on therapeutic interventions and the body's reaction to the therapeutics. The term *rampo* was used to describe medicine previously introduced by the Dutch. *Kampo* was used to differentiate it from *rampo*. *Kampo* is regulated by the Japanese government. Kampo is distinct from Chinese traditional medicine and Korean Traditional medicine.

Han Bang: Traditional Korean Medicine (TKM)

Han Bang is Korean Traditional Korean Medicine, and it is a fundamental part of the culture and used alongside "modern medicine". The life force, *ki*, like the Chinese *chi,* is the center of power for each living thing. It is responsible for all bodily functions, such as the circulation of blood and lymph, metabolism, and changes in our bodies as we age, and is our vital energy. *Ki* can be affected by the food we eat, illness, emotions and trauma. The key is to have a balanced *ki* and to store this *ki* in *danjun* (elixir field) to be circulated all over the body. Koreans believe there are 12 medians through which *ki* flows. Hang ban also incorporates Chinese traditional practices such as acupuncture, moxibustion, cupping and herbal remedies, which are described above.

Acupuncture is used to stimulate the flow of *ki.* It is used as a treatment for sprains, cramps and spasms, as well as many internal diseases and conditions. There are various forms acupuncture takes depending on what is being treated and where the flow of *ki* may be blocked.

For centuries in Korea, botanicals and herbs have been used to treat illnesses. A medical system evolved and was documented and published in a medical encyclopedia known as *Hyangyak Jipseongbang* (Classified Collection of Medical Prescriptions). This collection has over 703 native Korean medicines. It is a historic record of great value from the 15th century (published in 1433).

Ayurvedic Medicine

Ayurveda is a form of traditional medicine with an ancient lineage originating in India. The exact origin of Ayurveda is unknown, but it is believed to have been around for thousands of years. It is a system of traditional healing and preventive health care that has been practiced in India for centuries. It is thought to have existed since at least the 6th century BC. The earliest known written texts on Ayurveda were composed mostly during the Vedic period. These texts discuss various aspects of health and diseases, such as diet, lifestyle, anatomy, pharmacology, and pathology. Over the millennia, practitioners of this system of medicine have developed a comprehensive understanding and philosophy about health.

Ayurveda translates from Sanskrit to mean "the science of life" and is based on the principle that all living beings are made up of five elements – earth, water, fire, air, and ether – which form three fundamental energies known as doshas. These doshas are named Vata (air and ether), Pitta (fire and water), and Kapha (earth and water). Ayurveda is based on the belief that all health problems arise from an imbalance of the three doshas (or life forces). According to Ayurvedic theory, restoring the balance between these elements helps treat the underlying cause of the disease. Maintaining a balance between these doshas is thought to be essential to good health. This form of traditional medicine relies heavily on herbal remedies, massage therapy, meditation, and other holistic treatments to promote overall well-being.

African Traditional Medicine (ATM)

Since Africa is the cradle of humankind, it may be safe to say that African Traditional Medicine (ATM) is the oldest form of traditional medicine on the planet. Since Africa is a continent, ATM is practiced in differing forms to promote healthy lifestyles and healing. It has had a major impact on modern-day practice across the continent and is often used in conjunction with "modern" medicine. ATM traces its roots back to the earliest civilizations on the continent and has influenced many Western medical practices.

African traditional medicine uses a variety of techniques, such as herbal remedies, spiritual healing, massage therapy, bone setting, surgeries (including cranial surgeries which have been documented) and more, to diagnose and treat ailments to body, mind and soul. This knowledge is passed down through generations, eventually becoming an intrinsic part of many African cultures and belief systems. Traditional healers are still common in many parts of Africa today and play an important role in primary healthcare.

African plant-based botanicals and herbal remedies have been used for millennia to treat a variety of ailments and illnesses. Referred to as "traditional medicine" or "indigenous healing", the practice of utilizing plant-based therapeutics for medicinal purposes has its origins in ancient African cultures. Plant-based therapies are traditionally relied upon as primary sources of healthcare in many parts of the world, including Africa. Unlike modern medicine, these remedies are often derived from plants, roots, and trees that can be found in nature or grown by the practitioner.

African traditional medicine is centuries old and has its roots in ancient African healing practices. The use of herbs for medicinal purposes was first documented by Arabian physicians who wrote about their observations during explorations to West Africa. Traditional healers were responsible for providing healthcare and treatments to members of their communities, along with offering spiritual guidance. During this period, healers would often use herbs in combination with other natural remedies, such as massage and prayer, to treat illnesses.

Ancient Greece and Hippocrates

In Europe, traditional medicine has a long history. During ancient Greek times, Hippocrates wrote extensively about illness and its treatment using natural methods. He also believed in honoring individual differences when it came to healing - which is still very much a part of modern-day traditional medicine practices today. Other historical sources, such as the works of Pliny the Elder, Galen, and Dioscorides, also contain detailed descriptions of medicinal plants and their uses.

In conclusion to this chapter, it is safe to say traditional medicine continues to be a popular form of healthcare for many around the world. Its holistic approach to wellness has proven successful in treating numerous illnesses as well as promoting general health and well-being. From ancient Chinese texts to Greek sources, these timeless healing practices have been used throughout the ages with good reason.

Traditional medicine has a long and interesting history that spans centuries. From its origins in ancient China and India to its development in Greece, this branch of medicine has been able to offer effective treatments for many illnesses throughout history. As we move further into the future, it's likely that traditional medicine will continue to be an important part of our medical system, helping us maintain a healthy and balanced lifestyle.

This chapter provides a brief overview of the long and extensive history of traditional medicine. Traditional medicine has different yet similar theoretical bases. These practices have evolved over time, weeding out those that did not work and imbibing those that did through experimentation, beliefs and divination. Many of these practices have stood the test of time for thousands of years.

Chapter 2: Healing Practices Across Africa

African traditional medicine is one of the oldest and most comprehensive medical systems in the world. For centuries, people in various African cultures have used herbal remedies to treat a wide range of ailments. These age-old treatments are steeped in tradition and remain an important part of life in many African countries today. In this chapter, we look at the history of herbal remedies in Africa and explore how these treatments are still used today. This chapter will outline some of the medicinal plants and herbs used for medicinal purposes and their role in African culture.

Africa's traditional healing practices continue to be widely used today, especially in rural areas where access to modern medicine is limited. Herbal remedies are often combined with more conventional treatments to treat a range of conditions, from minor ailments such as headaches and stomachaches to more serious illnesses like malaria. In some regions, traditional healers still play an important role in the healthcare system and even serve as primary care providers for many individuals living in remote locations. In addition, herbs are increasingly being used in the cosmetic industry, with some African countries producing and exporting natural products such as shea butter and aloe vera.

Traditional Medicines Across Africa

Africa is known for its rich and varied history of traditional medicine, passed down through generations to provide treatments for a variety of illnesses and ailments. This includes herbal remedies that have been used in many African cultures for centuries or even millennia, as well as other approaches such as spiritual healing.

In East African countries such as Kenya, Uganda, and Tanzania, plants like Neem tree bark are boiled together to create a bitter remedy said to treat fever or malaria. The abundant plant life provides a rich source of medicinal plants, while local knowledge and experience are built up over generations regarding which plants can be used as treatments for different conditions. Here, some herbal medicines are taken orally, while others are applied directly to the skin or mixed with other ingredients, such as oil or honey, to treat wounds or ulcers.

In Central Africa, there is also a great diversity of traditional medicines used. Herbal remedies are used for many purposes here, too, including treating pain and inflammation, improving digestion, reducing fever, and helping to prevent infection. Common ingredients in these remedies include ginger root, turmeric powder, and aniseed – all known for their healing properties. Spiritual healing is also highly valued in Central Africa, and some traditional healers perform rituals such as dancing, chanting, and singing to help patients overcome physical or mental illness.

In West African countries like Nigeria and Ghana, natural ingredients such as leaves and roots are brewed into teas or mixed into topical ointments and poultices. Herbal remedies for a variety of ailments like fever, pain, and infection are still widely practiced today, often with prayers or incantations said as part of the treatment process. In addition to herbal medicines, spiritual elements such as sacrifices are sometimes included, too, depending on the context of each case.

Plant-based treatments are also popular in North African countries such as Morocco and Egypt, where popular herbs like sage, chamomile, mint and cloves are used to make teas that can be taken for cold symptoms. Here, herbs are not only used for their medicinal properties but also for spiritual purposes such as protection from evil spirits. Some of the plants commonly used include aloe vera, devil's claw, and dragon's blood – all known for their healing qualities.

In Southern Africa, there is a vast flora that has been used by traditional practitioners for thousands of years. People in this region and beyond have relied on this true and tested wealth of knowledge passed down through generations of healers in many different tribes and clans. ATM is used to activate the immune system and for cleansing and detoxifying the body. Other herbal medicine is used to stimulate the body to self-heal and/ or to combat symptoms. Prayer and sacrifice are also both common elements in Southern African traditional medicine as it is across most of Africa.

Types of Herbs Used for Medicinal Purposes in Africa

Herbal remedies have been used as traditional medicines in Africa for centuries, with different herbs and plants playing an important role in the treatment of various ailments and illnesses. For example, Aloe Vera has long been known to be a powerful antiseptic that can help heal any number of common skin conditions. The bitter herb Neem is also widely used to treat different types of fever, malaria, and other infectious diseases, while Turmeric is often taken to reduce inflammation and support wound healing. Other popular herbs include Garlic for its antiviral properties, Ginger for digestive issues, and Saffron for female reproductive health.

In addition to these health-promoting herbs, there are many other plants and herbs that have been used in Africa for centuries for their medicinal properties. The African Moringa tree is widely known for its anti-inflammatory benefits, as well as its ability to help boost the immune system. Similarly, Hibiscus tea has been used to reduce blood pressure and cholesterol levels and improve liver function. Other lesser-known but equally powerful medicinal plants include buchu leaves, which can help with urinary conditions; devil's claw, a powerful anti-inflammatory herb; and the adaptogenic herb ashwagandha, which can aid in stress relief.

Preparations and Applications

Herbal remedies can be prepared in a variety of ways, depending on the type of herb and its intended use. The most common methods include teas, decoctions (boiling down herbs in water), infusions (letting herbs sit in hot water to steep), and tinctures (infusing alcohol with herbs). When preparing an herbal remedy, it is important to follow the directions carefully, as different herbs require different amounts of time for preparation.

Once prepared, herbal remedies can be taken internally or externally, depending on the ailment being treated. For instance, tea made from chamomile flowers could be drunk as an enjoyable way to reduce stress or anxiety. Or a garlic-ginger infusion could be drunk to combat cold symptoms and strengthen the immune system. Meanwhile, tinctures such as echinacea can be applied topically to reduce inflammation and soothe skin conditions.

Herbal remedies are a simple and effective way to promote health and well-being, but it's important to remember that these treatments should only be used under the guidance of an experienced herbal practitioner or doctor. As with any form of medicine, it's best to consult a professional before taking any herbal remedy to ensure safety and effectiveness. By understanding the type of herbs available and their intended use, you can learn how to incorporate herbs into your daily life for improved health and overall well-being.

Why Do People Still Use Herbal Medicine Today?

Herbal remedies have been used for centuries all over the world, and Africa is no exception. Herbs are still widely used today in many parts of the continent, not only for their medicinal properties but also for spiritual purposes such as protection from evil spirits. This practice has endured despite the emergence of modern medicine and pharmaceuticals because of its proven efficacy and accessibility to communities that may lack access to more traditional forms of healthcare.

One reason why herbal remedies remain popular today is that they tend to be affordable and accessible, even in rural areas where healthcare is often limited or nonexistent. Many African countries have a shortage of resources when it comes to accessing medical advice and treatments, so using herbs can provide a cost-effective solution for illnesses and ailments. Additionally, herbs can often be sourced locally and prepared with minimal fuss, meaning that they are an easily accessible way of treating minor illnesses without the need to seek help from traditional medical providers.

Another reason why herbal remedies have endured is that they are often perceived as being more natural and less intrusive than modern medical treatments. Many Africans still hold traditional beliefs about health, so using a natural remedy such as herbs can feel safer and more culturally appropriate for people who may see modern drugs and treatments as unnatural or even invasive. For many, using herbs to treat illnesses is a way of connecting with their ancestral roots and tapping into the wisdom of traditional practices passed down from generation to generation.

Finally, it's important not to overlook the potential healing properties of herbs themselves. Many African nations boast an array of medicinal plants with powerful anti-inflammatory, immune-boosting, and adaptogenic qualities, which cannot be overlooked in terms of their effectiveness in treating various ailments. This is why herbal remedies have remained so popular; even with the emergence of modern medicine, people are increasingly recognizing that natural solutions can be just as effective – if not more so - than pharmaceuticals.

Although herbal remedies are a popular choice, they can also present certain challenges. Due to the lack of regulation in many African countries, it is difficult to assess the quality or safety of these treatments as there is no standardized testing process. Furthermore, there has been very little research done on the efficacy of traditional medicines compared to modern pharmaceuticals. Lastly, some traditional healers may lack the scientific knowledge and education needed to accurately diagnose and treat illnesses, making their treatments more dangerous than conventional ones.

By understanding the type of herbs available and their intended use, you can learn how to incorporate herbs into your daily life for improved health and overall well-being. Herbal remedies offer a safe and cost-effective way of treating minor illnesses while still allowing people to connect with their cultural roots. However, it is important to remember that they should always be taken under the guidance of an experienced professional. With the correct knowledge and understanding, herbal remedies can help people stay healthy without sacrificing cultural identity.

Chapter 3: Ayurveda

Ayurveda is an ancient healing tradition that has been used for thousands of years in India and other parts of the world. It is a holistic approach to health and wellness that focuses on restoring balance in the body, mind, and spirit. In this chapter, we will take a closer look at Ayurvedic principles and how they can be applied to live a healthier lifestyle. We'll explore various treatments, herbs, diet plans, and more to help you make positive changes in your life. Let's begin our journey into the fascinating world of Ayurveda.

Ayurveda is an integrative approach to health that views every individual as unique and recognizes imbalances in each person's energy as a cause of disease. Ayurvedic practitioners believe that illness is caused by blockages or imbalances in these energies and focus on restoring balance by creating dietary, lifestyle, and herbal remedies tailored to the individual. This holistic approach includes not just physical but also mental, emotional, and spiritual aspects of well-being.

Ayurvedic medicine is closely tied to yoga, and both are regarded as part of a unified system. In traditional Ayurvedic practice, these two disciplines work hand in hand and complement each other. For example, yoga helps maintain a balance between the doshas, while Ayurveda uses herbs and diet modifications to restore equilibrium if needed.

The main goal of Ayurvedic practice is to restore balance to prevent disease and promote longevity. It emphasizes lifestyle modifications and uses herbal remedies, diet plans, aromatherapy, massage therapy, yoga, meditation, and other therapeutic techniques that are designed to strengthen the body's own healing power. Proper nutrition is also important for good health, according to Ayurveda; certain foods can be used for treatment, and others should be avoided due to their potential side effects.

Ayurveda has gained popularity in recent years and is used by millions of people around the world. Its holistic approach to health and wellness has helped countless individuals achieve a better balance in their lives. Whether you're looking for natural remedies or just want to learn more about Ayurveda, this ancient practice is sure to provide beneficial insight into how you can lead a healthier life.

Treatment Options in Ayurveda

There are many treatment options available in Ayurveda, all of which are designed to bring balance to the body. Treatments can include herbal remedies, diet and lifestyle modifications, massage therapy, yoga and meditation, aromatherapy, and other natural healing techniques. Some of the most common treatments include:

Herbal Remedies, Spices, Plants, and Oils

Herbal remedies, spices, plants, and oils are some of the oldest forms of natural healing and remain popular today. In Ayurveda, herbs are used to create a balance between the body's three doshas – vata, pitta, and kapha. Herbal remedies can be administered in many different forms, such as tinctures, teas, decoctions (boiled herbs), powders, or capsules.

Some common Ayurvedic herbal remedies include ashwagandha (Indian ginseng), which helps reduce stress and is believed to increase energy levels; Brahmi, which supports cognitive function; Guggul for cholesterol management; Shatavari for women's health; and Turmeric which is known for its powerful anti-inflammatory properties.

Spices are an important component of Ayurveda, as they are used to balance the doshas and combat stress. Common spices such as black pepper, cinnamon, cumin, ginger, nutmeg, cardamom, fennel seeds, cloves, garlic, and turmeric can be used in herbal remedies or added to food for medicinal purposes. Plants may also be included in treatments such as aloe vera for skin health and neem for digestive issues. Oils can also be used therapeutically with a massage or aromatherapy using essential oils like lavender oil to reduce anxiety or eucalyptus oil to support respiratory health.

Diet and Lifestyle Modifications

Ayurveda emphasizes the importance of diet in maintaining overall health and wellness. Eating certain foods can support balance between the doshas, while others can create an imbalance. Ayurvedic practitioners generally recommend eating fresh, whole foods that are local, seasonal, and as natural as possible. Foods can also be cooked according to one's specific dosha type for optimal digestive health. Avoiding processed foods, caffeine, alcohol, fried food, white flour products, and animal products is recommended to maintain a healthy balance within the body.

In addition to following a balanced diet, lifestyle modifications are important in Ayurvedic practices. Practitioners recommend getting plenty of rest and exercise, including yoga or other forms of physical activity. Meditating daily can help bring balance to the mind, body, and spirit. Other lifestyle modifications include avoiding negative thoughts and actions, focusing on relationships with family and friends, and engaging in meaningful activities that make life more enjoyable.

Massage Therapy

Ayurveda considers massage an important part of healing due to its ability to move energy through the body. Massage therapies such as abhyanga (whole-body oil massage) can help relax muscles, reduce tension, improve circulation, boost immunity, and support overall health. Different types of oils are used based on one's dosha type - sesame oil for vata, sunflower oil for pitta, and coconut oil for kapha.

Ayurvedic massage also uses marma points, which are energy points on the body that can be manipulated to enhance health and well-being. Marma point massage is a form of deep tissue therapy that works to unblock stuck energy throughout the body and promote balance. Ayurvedic practitioners may use different techniques, such as rocking, kneading, tapping, vibrating, and stroking during massage therapy sessions.

Ayurvedic self-massage is another option for those looking to practice at home. Self-massage helps facilitate lymphatic drainage, reduce stress levels and tension in muscles, aid digestion issues, improve circulation, and calm the mind. During self-massage, you should begin with your feet, starting at the toes and working your way up to the thighs. Use a warm oil of your choice to massage each area for several minutes before wiping off excess oil with a dry cloth or towel. Working through all areas of the body should take around 20-30 minutes.

Yoga and Meditation

Yoga is an important form of exercise in Ayurveda as it helps to balance the body's three doshas while increasing flexibility and strength. Different yoga poses are used depending on one's individual needs, with a focus on proper breathing techniques. Yoga can be helpful for those who are looking to reduce stress and increase energy levels. It also helps to improve concentration, focus, and balance the nervous system. Practicing yoga on a regular basis can help relax the mind and body while improving overall well-being.

Meditation is also a key part of Ayurveda, with various techniques such as vipassana (mindfulness) or mantra (chanting) designed to reduce stress levels, improve concentration, and bring clarity to one's life. Meditation can help to cultivate a sense of inner peace and balance, which is essential for overall health. Focusing on the breath is an important part of meditation and can help to bring clarity and stillness to the mind.

When practicing Ayurveda, one should also take time for self-care through activities such as taking a warm bath or listening to calming music. Engaging in creative activities such as drawing, dancing, writing, and cooking can also be beneficial for reducing stress levels and improving inner peace.

Aromatherapy

Aromatherapy is an integral part of Ayurveda and can be used to uplift one's mood, reduce stress levels, and improve the general sense of well-being. Essential oils are extracted from plants through steam distillation and contain concentrated extracts of herbs and flowers with healing properties. Various oils have different effects on the body, such as sandalwood for calming, lavender for reducing anxiety, geranium for relaxation, lemon grass for energizing, and so on.

Ayurvedic aromatherapy works by inhaling essential oils directly or diffusing them in a room. Inhalation allows the molecules from the oils to enter your bloodstream via the respiratory system, while diffusion helps to spread the scent throughout the room. Aromatherapy can also be used in massage therapy by adding a few drops of essential oil to natural carrier oils such as jojoba, sesame, and coconut.

When using aromatherapy, it is important to choose the right oil for your individual needs and any underlying health conditions you may have. Always consult a qualified practitioner first before using essential oils to ensure maximum safety and effectiveness. It is also important to note that some oils should not be used during pregnancy or if you suffer from asthma or allergies.

Influence of Ayurveda on other Medical Practices

Ayurveda has been influencing and shaping medical traditions around the world for centuries. From *Unani Tibb* to *Siddha Vaidya* in

India to its integration into mainstream modern medicine today, Ayurveda has had a far-reaching impact on health care globally.

Unani Tibb

Unani Tibb is a holistic healing system that originated in Greece but was heavily influenced by Ayurvedic principles. It seeks to restore the balance between the body's four humors: black bile, yellow bile, phlegm, and blood. Like Ayurveda, it emphasizes strengthening the body's natural healing powers, using diet and lifestyle modifications as well as herbal remedies. It is believed that Unani Tibb was introduced to India by the Arabs and eventually combined with Indian Ayurvedic medicine to form a unique system of healing known as Siddha Vaidya.

Siddha Vaidya

Siddha Vaidya is a blend of both Ayurveda and Unani Tibb, focusing on preventing illness by maintaining harmony between the mind, body, and environment. It uses several methods, such as cleansing therapies, special diets, detoxification procedures, massage therapy, yoga, and meditation, to restore balance within the individual. Herbal remedies are also used to treat specific ailments.

Uses In Modern Medicine

Ayurveda has been gaining more popularity in recent years and is increasingly being integrated into modern Western medicine. The World Health Organization (WHO) recognizes Ayurveda as a traditional system of medicine with the potential to address modern health issues such as stress, chronic pain, fatigue, and digestive disorders. Many hospitals and clinics around the world are beginning to incorporate Ayurvedic treatments into their patient care plans as well as offering special courses on Ayurvedic healing protocols for medical professionals.

For example, some studies have shown that Ayurvedic massage therapies may be beneficial for reducing inflammation and improving joint mobility in patients with arthritis or chronic musculoskeletal pain. Certain herbs and supplements have also been found to be effective in reducing high cholesterol levels and preventing certain diseases, such as cancer, while others may help relieve symptoms of anxiety or depression.

Ayurveda's holistic approach to health care focuses on the individual, considering their physical, mental, emotional, and spiritual needs. This makes it an ideal complement to traditional Western medicine by providing a more personalized approach that can effectively target underlying imbalances in the body. With its long history of success and growing acceptance within the medical community, Ayurveda is sure to continue influencing medical practices around the world for many years to come.

In conclusion, Ayurveda is an ancient medical practice designed to promote holistic wellness through various means, such as massage therapy, yoga and meditation, self-care activities, and aromatherapy. By incorporating the principles of Ayurveda into your daily life, you can bring balance and harmony to both mind and body. With dedication, patience, and consistency, you can reap the many benefits of this natural healing system on your own health journey.

Chapter 4: Traditional Chinese Medicine

Traditional Chinese medicine (TCM) is a holistic approach to health and well-being that has been used for over two thousand years. It encompasses many healing modalities, such as acupuncture, moxibustion, cupping, herbal remedies, and Qi Gong exercises. TCM is based on the idea of balancing the body's energies to achieve optimum wellness. By restoring balance, practitioners of TCM can aid individuals in relieving physical and emotional ailments, ranging from common colds to chronic medical conditions.

In this chapter, we will explore the history of TCM and examine how it is still practiced today, both in China and around the world. We will look at each modality individually to gain an understanding of how it works and examine the evidence for its effectiveness.

Today, TCM is practiced around the world by trained professionals who use tried-and-true methods that have been passed down through numerous generations. These include acupuncture, moxibustion, cupping, herbal remedies, massage therapy, and Qi Gong exercises. All these treatments are designed to restore balance to the body, which can help alleviate physical and emotional ailments.

Different Types of Traditional Practices

Traditional Chinese Medicine encompasses a wide range of therapies and treatments, each with its own unique benefits. Acupuncture, moxibustion, cupping therapy, herbal remedies, and Qi Gong exercises are all modalities of TCM that have been used for centuries to improve health and well-being.

Acupuncture

Acupuncture is an ancient form of Chinese medicine that involves the insertion of extremely thin needles into specific points of the body. This is believed to stimulate the body's own healing processes, restoring balance and promoting well-being. It is one of the most widely used forms of Traditional Chinese Medicine (TCM) and is practiced all over the world.

Today, acupuncture is widely used for a variety of ailments ranging from chronic pain to stress-related conditions like anxiety and depression. It is believed to be effective because the needles stimulate certain points on the body, known as "acupoints", which in turn activate various biochemical processes that can help reduce pain and improve overall well-being.

When it comes to evidence for effectiveness, acupuncture has been studied extensively over the past few decades. Numerous clinical trials have suggested that it can provide relief from chronic pain, including that caused by fibromyalgia and osteoarthritis. Other studies have also found positive benefits for conditions such as migraine headaches, menstrual cramps, stroke recovery, and mental health disorders like anxiety and depression.

The benefits of acupuncture go beyond just physical ailments, though; many people find that regular sessions can also help them relax and reduce stress. This is because the needles stimulate certain pathways in the brain, which can trigger a state of deep relaxation, allowing for better sleep quality and improved mood.

Like any medical treatment, acupuncture does have some risks associated with it. These include pain at the insertion site, bruising, bleeding, or infection if the needles are not properly sterilized. Despite these potential risks, though, acupuncture remains one of the most widely practiced forms of Traditional Chinese Medicine due to its proven effectiveness in treating a variety of physical and mental health conditions.

Moxibustion

Moxibustion is a form of Traditional Chinese Medicine (TCM) that predates acupuncture. It has been used with acupuncture for centuries to treat various ailments. It involves the burning of mugwort, a spongy, fluffy herb (Artemisia vulgaris) found in Asia, Africa and Europe. Burning mugwort can be done either directly on the skin, indirectly using a special stick, or by using an acupuncture needle to stimulate the flow of Qi and balance energy within the body.

Moxibustion works by stimulating specific acupoints with heat generated from burning the leaves, also known as moxa. This is believed to help restore balance and improve overall health by promoting better circulation and improved energy flow throughout the body. In some cases, moxibustion may also be used in combination with acupuncture to provide a more comprehensive treatment. Studies have found that it can be effective in relieving pain caused by arthritis as well as reducing symptoms of fatigue and depression. It has also been used successfully to treat infertility, morning sickness, and other gynecological conditions.

There are various types of moxibustion treatments available depending on the condition being treated. Direct moxibustion involves the burning of moxa sticks directly on the skin, while indirect moxibustion uses an acupuncture needle to transfer heat from the burning herb to specific acupoints. Other techniques include smokeless moxa and non-burning herbal plasters, which can be applied topically.

The benefits of moxibustion are numerous, including improved circulation and increased energy levels. It can also help reduce stress levels, alleviate pain, improve moods, and promote a balanced emotional state. In addition to these benefits, the use of moxibustion can also enhance the effectiveness of acupuncture treatments.

Some risks associated with moxibustion should be considered before undergoing treatment. These include potential skin irritation or burns, as well as possible side effects such as nausea or dizziness. However, when used correctly, these risks are minimal, and moxibustion is generally considered to be a safe form of therapy.

Cupping Therapy

Cupping therapy is an ancient practice used in traditional Chinese medicine (TCM). It relies on creating suction using cups made from glass, bamboo, or other materials. The therapist places the heated cup onto the patient's skin and applies pressure to create a vacuum. This vacuum creates suction that draws the underlying tissues up into the cup, resulting in increased blood flow to those areas.

Studies have reported that cupping therapy is effective in treating joint pain, neck and shoulder tension, and muscle spasms. Studies have also found that it is effective in treating other conditions, such as asthma, eczema, and digestive issues. It is believed to help reduce inflammation, stimulate healing, and improve overall well-being.

There are three types of cupping treatments: dry cupping, wet cupping, and flash cupping. Dry cupping involves placing a heated cup onto the skin and allowing it to remain there for several minutes before removing it. Wet cupping includes pricking the skin with needles before applying the cup to draw out blood that has been blocked by stagnation or inflammation. Flash cupping requires very quick applications of cups over larger areas of skin at once.

The benefits of cupping therapy may include increased blood circulation to targeted areas and improved mobility of the affected muscles. It may also reduce pain, inflammation, and stress levels. However, cupping is not without risks. These include bruising, burning or reddening of the skin, and infection if done improperly or if proper hygiene practices are not followed.

Qi Gong Exercises

The practice of Qi Gong Exercises involves physical movements, such as stretching and certain postures that are meant to aid in balancing and harmonizing qi (energy) throughout the body. It is believed that these movements can help with the body's natural healing process by stimulating the flow of qi in specific areas. In addition to physical postures, Qi Gong practice includes breathing exercises and meditation, which are meant to bring about a state of relaxed focus.

A growing body of evidence suggests that Qi Gong exercises may be an effective way of relieving pain and improving overall well-being. For instance, a systematic review found that Qi Gong exercise may have positive effects on physical health, mental health, and quality of life for those experiencing various forms of chronic illness or injury. Other research has suggested that Qi Gong exercises may have a positive effect on reducing stress and anxiety, as well as improving cognitive performance.

Qi Gong exercises are divided into two main categories: dynamic and static. Dynamic exercises involve active movements such as postures, stretches, and breathing techniques meant to increase circulation and balance qi. Static exercises are focused on stillness and relaxation, with the goal of calming the mind and body while cultivating inner peace. Other types of Qi Gong exercises include medical Qi Gong, martial arts Qi Gong, and spiritual Qi Gong, which are often used in meditation practices.

Practicing Qi Gong exercises can offer several potential health benefits. These may include improving flexibility, muscular strength, and balance, reducing stress and anxiety, increasing energy levels, and improving the overall quality of life. However, it is important to note that there are also potential risks associated with practicing Qi Gong, such as injury from improper form or overly vigorous movements. In general, it is recommended to consult a healthcare professional before beginning any new exercise program.

Traditional Chinese medicine is an ancient system of healthcare that has endured for centuries and is still widely practiced today. It is based on the concept of balance in terms of mind and body, as well as yin-yang forces. Through a variety of treatments such as acupuncture, moxibustion, cupping, herbal remedies, and Qi Gong exercises, it seeks to restore harmony within the body by unblocking energy pathways, promoting relaxation, relieving pain, and improving overall health. In this way, TCM successfully provides holistic care for patients all over the world.

As modern science continues to learn more about TCM's effectiveness and its potential applications in Western medicine, this ancient practice may yet reveal new ways of treating various ailments. Whatever the future holds, traditional Chinese medicine will remain a popular and powerful healthcare system for many years to come.

Chapter 5: The Golden Age of Medicine from The Near East

The Near East region of the world has a long history of medical practice and advances that began during the Golden Age. From the empires of Ottoman and Persian to the Arabian Peninsula, a unique form of medicine developed which was based on different beliefs and practices than those in Europe at the time. This chapter will explore some of these advances made during this period and how they still influence modern-day medicine.

Medical Advances During the Golden Age

The Near East region of the world had a long tradition of medical practice, many of which were adopted from even earlier times, such as Ancient Egypt and Babylon. During the golden age, advances in medicine took off across multiple empires and cities such as Istanbul, Damascus, Baghdad, and Cairo. These advances included hygiene techniques to keep disease at bay and promote health; pharmacology advancement with knowledge of herbs, spices, and minerals; use of chemical medicines to treat infections; developments in surgical instruments and practices; and new treatments for mental illness.

The advances made in medicine during the golden age of the Near East ranged from simple techniques for disease prevention to more complex treatments. Hygiene was a major focus, with research and practice into the use of bathing and washing hands to reduce infection rates. In addition, pharmacology had its own advances as those living during this time sought out herbs and spices to treat various ailments. Compounds such as turmeric, cloves, mint, saffron, ginger, garlic, cinnamon, and myrrh were all used in remedies for common illnesses.

Surgical advancements also took place during this era with discoveries in methods for the treatment of fractures and dislocations, among other procedures. This included understanding anatomy better as well as the development of specific tools to aid in surgery, such as the lancet, speculum, and retractor. Mental health was also explored during this period with the development of treatments like music therapy, aromatherapy, and hypnosis.

The advances made during the golden age of medicine in the Near East region have had a lasting influence on modern-day medicine. Many of these techniques are still practiced today, such as hygiene and hand washing methods to prevent infections, herbal remedies for treating illnesses, surgical interventions, and mental health treatments. As such, it is important to recognize that many modern-day advancements have their roots in this period of history and should be acknowledged accordingly. A greater understanding of how the practices used by early civilizations can help inform our current medical practices and aid in finding new discoveries.

Chemists and Scholars

Chemists and scholars played an important role during the Golden Age of medicine in the Near East. This period saw a surge in scientific advancement through research, experimentation, and observation, which led to new discoveries about human anatomy, diseases, and treatments. Chemists developed formulas for medicines and tinctures that were effective against various ailments, while scholars studied medical texts from ancient civilizations such as China, India, Greece, Persia, and Egypt to gain new insights into medicine.

The availability of vast archives of medical information made it possible for minds like Avicenna (Ibn Sina) to develop theories regarding modern-day treatments such as antibiotics. He was one of the earliest chemists to use herbs and plants as natural remedies for a variety of illnesses, such as fever and headaches. His book Canon of Medicine was used by medical students in universities across the Near East region.

Another notable scholar during this period was Abu al-Qasim al-Zahrawi (Albucasis), who studied Greek medicine extensively before writing his own comprehensive treatise on surgery known as Al-Tasrif. This work contained detailed descriptions of different surgical techniques that are still relevant today, and it has served as an important reference for physicians around the world.

Chemists and scholars during the Golden Age of medicine in the near east region contributed greatly to modern-day treatments. Their discoveries and advancements continue to be studied today, offering insight into how medical science has evolved over time. By understanding the work of these pioneer minds, current researchers can build upon their foundations to create even more effective treatments for the benefit of all humanity.

Hygienic Practices

Hygienic practices were a major focus of medical advances during the Golden Age in the near east. Ancient civilizations such as those from Egypt and Babylon had already developed concepts of hygiene and sanitation, but these practices were further explored during this period. The various empires of the region sought to reduce infection rates and promote health through techniques such as bathing, washing hands, deodorizing rooms with incense or herbs, disposing of waste properly, and limiting contact with animals that could spread disease.

Bathing was seen as an important practice for physical and spiritual purification which was adopted throughout much of the region. People would take baths before any type of religious ceremony or celebration as well as for health reasons. Baths could range from simple pools of water to public baths with heated floors and steam rooms, depending on the region and culture.

Hand washing was another important practice during the golden age of medicine in the near east, especially for those working in medical settings. Medical professionals were instructed to wash their hands with a mixture of water and soap before touching any patient or handling any tools used for surgery. This technique was also adopted by many households that would use a combination of water, lemon juice, and salt as a disinfectant.

The importance placed on hygiene during this period extended beyond just physical cleanliness; practices such as deodorizing rooms with incense or herbs were also adopted to make spaces more pleasant. Disposal of waste and limiting contact with animals, especially in areas where diseases such as malaria were common, was also practiced, reducing infection rates.

Overall, hygienic practices played an important role in the medical advances made during the golden age of medicine in the near east region. Such practices are still relevant today and have helped to improve overall health and longevity for all people. By learning from these ancient cultures, modern societies can continue to develop effective hygiene techniques that can benefit everyone.

Other Medical Breakthroughs

It is evident the role hygiene plays in mitigating the spread of disease. In 1854, approximately 700 people died within a few weeks. Cholera, known as “the Blue Death,” was running rampant in London. It was the English physician John Snow, that tracked the outbreak to a certain water pump on Broad Street in the Soho district of London. This hypothesis was followed by Reverend Whitehead who at first disputed Dr. Snow’s hypothesis. Once the handle of the pump was inoperable, the cases of the disease declined.

The Golden Age of medicine in the near east also saw several notable medical breakthroughs that are still relevant today. This period saw advances in the use of herbs and spices for medicinal purposes, an increase in knowledge about mental disorders, and the development of new surgical techniques.

Mental health was also explored during this period; physicians such as Rhazes (Rhases) from Persia are credited with recognizing mental illness and developing concepts to distinguish between physical ailments, demonic possession, and psychological disturbances. This has helped pave the way for modern-day treatments for mental health disorders.

The use of leeches in medicine was another medical breakthrough that originated during this period. Leeches were used to treat ailments such as headaches, gout, and various skin diseases, as well as to reduce swelling from wounds or bruises. The practice of bloodletting was also common during this time, although it has since fallen out of favor due to its potentially dangerous side effects.

In addition, Arab physicians developed an early form of vaccination for smallpox, which involved taking material from a patient who had recovered from the disease and then introducing it into the body of another person to create immunity against infection.

The Golden Age of medicine in the near east was a period of significant advancement and discovery, with contributions from chemists, scholars, and physicians that continue to shape modern-day healthcare. During this time, powerful antibiotics were developed, sanitation methods improved, leeches were used for medical purposes, and vaccinations were introduced as preventative measures against diseases like smallpox.

The work of these pioneering minds during this period has had an incredible impact on modern-day medicine. By studying their research and understanding traditional remedies, researchers can build upon their foundations to create even more effective treatments for the benefit of all humanity. The contributions made by medical practitioners during the Golden Age of medicine in the near east are worth remembering and celebrating.

Chapter 6: The Americas, Traditional Shamanistic Medicine

In the Americas, just like in Asia, Africa and Europe, traditional medicines include a rich diversity of cultural and traditional beliefs, practices, and remedies. Traditional medicinal practices are passed down from one generation to another in indigenous communities, as seen across the world. There is almost always a deep connection to the environment, to nature, the animals and trees, as well as other plants. In addition, there are usually spiritual elements and elements of worship. As with all traditional medicine and healing practices, people who seek out alternative medicine need to do so with respect for the cultures, the practices and the practitioners.

North America

In the United States of America and Canada, traditional medicine is practiced by many Indigenous American Tribes. These tribes possess a deep knowledge of the medicinal properties of the natural resources specific to the region. Some of these, such as cedar, sage and sweetgrass, are used in ceremonies for purification. These are common herbs used in purification rituals across the world in other indigenous cultures.

In Canada, the term *First Nations* is used for the indigenous peoples who fall under the umbrella of 600 recognized First Nations. These tribes have their own distinct histories, languages and cultures. The *Inuit* live in the Arctic regions, including Labrador, Nanavut, Northwest Territories, and Quebec. They have a very distinct way of life and culture, such as building igloos and dog sledding. Métis are a mixed race of indigenous and European ancestry. They mainly live on the Prairies in the regions of Alberta, Manitoba and Saskatchewan.

Other Alaskan indigenous people fall under the categories of the *Aleut*, *Athabaskan*, *Haida*, *Iñupiat*, *Tlingit*, *Tsimshian* and the *Yup'ik*. Within these indigenous groups, there are also smaller communities and nations with their own identities. It is important to note that there is a rich diversity of multiple nations, but the purpose of this is to illustrate that traditional medicines and cultural practices are performed in many different ways across this part of the world, and only a few are mentioned in this chapter.

Medicinal Plants

First Nations and other Native American tribes have a deep knowledge of the ecosystems in their environment. They take a holistic approach to life, which includes the importance of the environment, the animals, and the flora. Indigenous First Nation healers use plants such as Bearberry leaves to make tea that has healing properties, is a diuretic and helps with urinary tract problems. Oplopanax horridus, Devil's Club, has large spiky leaves and many First Nations believe it has anti-inflammatory properties as well as being a plant that boosts the immune system. Juniper berries are known worldwide for medicinal uses. Some plants with antibiotic properties, such as Rhododendron Groenlandicum, known as Labrador Tea, are used to treat respiratory issues. Asarum Canadense, also known as Wild Ginger, is used in teas for its anti-inflammatory properties. Yarrow is often used as a poultice and is believed to have properties that accelerate the healing of wounds when used in a poultice.

Sweat Lodges.

Inipi is a Lakota Sioux word from the Great Plains in North America. This word means "to live again" or to be "born again". This is a traditional practice that many Native American tribes use as purification rites that renew a person's or people's lives through cleansing / purification. This is done in conjunction with intentions, prayers and or chants for spiritual guidance.

There are many variations of sweat lodge ceremonies across the different Native American tribes. Each tribe has developed its own *inipi* traditions according to tribal practices. In Mexico, the *Temazcal* is the name given to the sweat lodge ceremony, and the practitioners are known as *Temazcalero* or *Temazcalera.*

The physical structure where *inipi* takes place is usually an enclosed circular space where heat, steam and smoke are trapped. Ancestral or grandfather stones made from volcanic rock (also known as "spirit stones") are heated in a fire outside and brought into the sweat lodge. It is here that people gather to pray and sing songs and chants. Water with purifying herbs is often poured onto the stones to create steam and heat, which are believed to have purifying properties. This is believed to provide a deeper connection to the realm of the ancestors and spirits. This ritual is believed to cleanse those who take part in it to create a rebirth. Most of the time, this is done in a community with others and brings people closer together as they go through this experience, which is physical, spiritual and often also emotional.

In Mexico, there are several names for traditional healers: *Curandero* (male) or *Curandera* (female) use a traditional, holistic approach to healing that includes herbs and spirituality, including rituals, prayers and fasting to assist the healing process. Herbalists in Mexico are often referred to as *Yerbero* or *Yerbatero* and specialize in the use of plants that have medicinal properties. They prepare tinctures, poultices and teas for many health conditions. Traditional bone setters are known as *Huesero* or *Huesera.* They treat musculoskeletal issues such as fractures with manipulation and poultices.

The term used for traditional massage in Mexico is *masaje tradicional.* Like Bonesetters, the traditional massage therapist often gains generational knowledge that is passed down through family lineages. Traditional massage therapists perform manual manipulation of joints and muscles and are known as *Sobador* (male) or *Sobadora* (female). There are specific techniques used to promote relaxation and enhance circulation. However, they are often called upon to help with the alleviation of pain in the joints or other musculoskeletal issues.

Central America

Amongst the Maya in Belize, El Salvador, Guatemala, Honduras and Mexico, traditional healers or shamans are called *Ajq'ij*. This means Daykeeper, which is how the Maya refer to their spiritual healers. The Ajq'il are important in Maya traditional lifestyles, where they are involved in healing and creating harmony with the natural and spiritual worlds. The Ajq'ij are also leaders in rituals, ceremonies, and divination and are knowledgeable when it comes to the Maya solar calendar systems. The calendar systems of the Maya have two separate calendars. One is the *Tzolkin,* which is the sacred calendar, and the other is the *Haab*, the solar calendar.

Caribbean

In the Caribbean, due to the history of the merging of multiple peoples and cultures, Indigenous, African, Asian, and European religious beliefs, traditions and healing have all played a role in the way Traditional Caribbean medicine is practiced. Like many other regions of the world, traditional medicine is passed down orally from one generation to the next. The elders who practice traditional medicine are pivotal in training future generations in the techniques used, the herbal remedies and practices.

Many rituals in the Caribbean are associated with traditional medicine and traditional healing. These are influenced by cultural and religious practices and are often communal in nature through ceremonial healing. This is done for physical and spiritual imbalances, and often, the whole community takes part, which builds communal cohesiveness.

Efforts are underway to incorporate traditional medicines into the broader healthcare system while still maintaining respect for the diverse cultural heritage. Practices differ across the region. However, one thing they all have in common is the historical influences.

"Bush Medicine" is the term used for herbs, spices, bark of trees, leaves and other local resources used in healing ailments. Medicinal plants are commonly found in the tropical environment and used for treating ailments, and to promote general well-being both physical and spiritual.

Obeah is often thought of as a spiritual and magical practice with roots in Indigenous traditions, West African spiritualism and European occultism. It is a term that is used also by the Caribbean diaspora community in North America. It is practiced by Obeah Workers or spiritual workers, sometimes known as *Obeah* "doctors". Obeah differs from Voodoo as it does not have structured deities and is less structured among practitioners. Many who practice have suffered discrimination and prejudice. Some practitioners have been linked to criminal activities, which adds to the stigmatization.

Voodoo or Vodun (as practiced in parts of Benin, Togo and Nigeria) was brought over to the Caribbean by people through the Atlantic Slave Trade. It is also practiced in Central America, Haiti, and Louisiana in the United States. Voodoo priests, *Houngan,* and priestesses, *Mambo*, (in Haitian Vodou), are known for their role in community ceremonies and serve in various capacities, such as having responsibilities for healing rituals (physical and spiritual). Energy healing is an important part of Voodoo practice.

Traditional medicines used in *Voodoo*, like most herbal medicines across the world, are taken in the form of teas or infusions, poultices, and ointments applied topically. Tree bark and roots of plants are used in traditional medicine practices, passed down from generation to generation, with a deep shared heritage in West Africa. Ceremonial cleansings occur in the form of baths that are infused with herbs, bark and other naturally occurring substances. These rituals invoke positive energies through chants and drumming.

Ceremonial cleansing uses herbal-infused water and herbal and bark baths for ritual cleansing to drive away negative influences and deliver positive energy. Typically, herbs used are basil, bay leaves, cinnamon, eucalyptus, mint, rosemary, and cascarilla, which are made from ground eggshells.

South America

In South America, traditional medicine varies with the different indigenous communities. These practices are gaining acceptance and recognition for their contributions to healthcare and are integrated into current-day approaches.

Andean Traditional Medicine, as practiced in Peru and Bolivia, involves Erythroxylaceae, the shrub on which coca leaves grow. The coca leaf is a known stimulant and has been used for centuries due to its medicinal properties. It grows in Argentina, Bolivia, Columbia, Ecuador and Peru. The coca leaf is chewed and used in teas, which helps to alleviate the effects of altitude sickness. It also helps to alleviate hunger and increases energy.

Despacho originates from the Spanish word that means to dispatch. The Despacho Ceremony refers to an offering made up of herbs, flowers, seeds and sweets. It is a ritual that involves symbolic offerings for healing and balance for mind and body.

Amazon Rainforest Traditional Medicine practitioners utilize some of the following plants:

Ayahuasca is a plant mixture consumed during ceremonies by indigenous tribes of the Amazon, such as the *Shipibo-Conibo* and *Asháninka. Ayahuasca* vine (*Banisteriopsis caapi*) is mixed with the leaves of the *chacruna* plant (*Psychotria viridis*) into a hallucinogenic brew. The mixture can affect one's mind, mood and or consciousness.

These psychoactive plant mixtures are used for spiritual and healing purposes but are usually hallucinogens, which can cause visual or auditory hallucinations or changes in consciousness. This plant mixture has the potential to lead to dependence. Psychotropic substances are heavily regulated and, in some regions, are banned. The potential risks, legal and health-related need to be weighed due to the implications of their use.

Medicinal Plants are used by the *Mapuche* people of Chile and Argentina by *machis* (healers), like *boldo* (*Peumus boldus*), which is an evergreen shrub or small tree. The leaves of the boldo plant have a strong and distinct smell. They are used to make teas that have medicinal properties and promote digestive health. It is also believed to protect the liver, have antioxidant properties, and guard against urinary tract infections.

Two other herbal medicines are *Açaí Berry* and *Euterpe Oleracea.* They are globally used superfoods that originated in the Americas. They have been a traditional source of food for Native Americans and are known for their antioxidant properties and are used to promote overall health. *Muirapuama, (Ptychopetalum olacoides)* is a traditional remedy that is prepared as a tea or a stronger solution in a tincture. It is used for curing fatigue and menstrual disorders.

In conclusion, as far as Traditional Medicines in the Americas is concerned, there are many and varied practices across the Americas, including as far north as the Arctic south across Alaska, Canada through Mexico, Nicaragua, and South America, along with Amazon. Thus, not only is it crucial to have knowledge of these traditional herbal substances but also to understand their effects and consequences. That is why it is highly recommended to seek professional guidance before using them.

Chapter 7: Natural Foods for Healing

Coming from different corners of the world, natural foods have been used for centuries to prevent and treat diseases. From ancient Chinese herbal medicine to Ayurvedic remedies, these plant-based therapies have been a source of healing for generations. In this chapter, we will look at the scientific basis behind using natural foods for healing, exploring their properties and uses from around the world.

Natural foods for healing have been used since ancient times, and cultures all over the world continue to practice this form of medicine. Natural foods are those that are derived from plants, animals, and other natural sources, as opposed to medicines or supplements that contain synthetic substances. The use of natural foods for healing is based on their unique properties, which provide a variety of medicinal benefits.

A definition of natural foods for healing is *any food item that has its own intrinsic medicinal properties*. These properties may be found in naturally occurring compounds such as vitamins, minerals, enzymes, and antioxidants, which can help the body fight off infection and promote better overall health. Many cultures have employed various combinations of natural ingredients in herbal remedies, teas, and tinctures to treat a wide range of ailments.

An overview of different cultures' use of natural foods for healing can be seen in the way they have used plants, animal products, and other natural sources to create remedies. In traditional Chinese medicine (TCM), practitioners have long utilized herbal medicines, including ginseng, garlic, ginger, and turmeric, to treat various ailments. Similarly, Ayurveda from India has prescribed many plant-based remedies such as neem leaves for skin problems or cumin for digestive issues. The Incas and Mayans employed coca leaves in their treatments, while African tribes boiled bitter roots to make an infusion that was believed to cure headaches and help prevent malaria. These examples give just a few insights into the vast range of natural remedies used around the world.

The scientific basis behind using natural foods for healing lies in the fact that they contain many beneficial properties that can be harnessed to promote health. For example, many herbs and spices have anti-inflammatory and antioxidant properties that can help fight off infection and reduce inflammation. Other plant-based items such as fruits, vegetables, seeds, and nuts provide essential vitamins, minerals, and fiber, which are vital for overall bodily functions. Finally, animal products like fish oil have omega-3 fatty acids, which are important for brain development and heart health. All these compounds work together to create a powerful form of medicine when used correctly.

Properties of Natural Foods

Superfoods and antioxidants are two of the most powerful natural foods for healing. Superfoods like blueberries and acai berries are high in antioxidants, which help to neutralize free radicals in the body and reduce inflammation. Additionally, these superfoods contain phytochemicals that can boost immunity and strengthen cells against attack from bacteria, viruses, and other pathogens. Vitamins, minerals, and phytochemicals found in natural foods also work together to nourish the body with the essential nutrients it needs to function optimally.

Vitamins such as A, B-complex, C, D, E, and K all have important functions in the body; for example, vitamin C helps to build collagen, while vitamins E and K are essential for blood clotting. Minerals such as calcium, magnesium, and zinc are integral to optimal health, helping to regulate the body's PH balance, enable proper muscle and nerve function, and support healthy bones. Phytochemicals found in plants have been linked to fighting cancer, reducing inflammation, improving cardiovascular health, boosting immunity, and providing several other beneficial effects.

Uses of Different Natural Foods

This section of the book contains natural foods that have been used traditionally for centuries to aid and improve physical and mental health. There is ample scientific evidence that shows eating certain foods or using them in teas or poultices, for example, can provide health benefits.

Ginger root and turmeric root have long been used for medicinal purposes, and their healing properties are now widely accepted. In Traditional Chinese Medicine (TCM), ginger is believed to increase digestive power, reduce inflammation, and relieve pain. TCM also suggests that ginger can help with heartburn, nausea, arthritis pains, and even cancer prevention. Ginger has also been used for centuries in Ayurveda for its anti-inflammatory and digestive properties. Similarly, it has been used to treat colds and flu, rheumatism, headaches, and other ailments.

Garlic has been used as a natural remedy since ancient times, and it is still widely used today due to its antiviral, antibacterial, and antifungal properties. Garlic can be eaten raw or cooked, crushed or chopped – whichever way you choose, the benefits are plentiful! The active component of garlic (allicin) helps fight off bacteria that cause infections such as bronchitis, tuberculosis, and pneumonia. It also reduces symptoms associated with colds, such as fever, chills, and sore throat. Additionally, research suggests that garlic may help lower cholesterol and protect against heart disease.

Turmeric is a spice that has long been used for its healing properties, and it is no surprise that it is still a popular remedy today. Curcumin, the active ingredient in turmeric, has been shown to have anti-inflammatory, antioxidant, and antiviral effects. Studies suggest that consuming curcumin may reduce inflammation caused by arthritis, decrease symptoms of depression, improve liver function, and even help prevent cancer.

Ginseng is an herbaceous perennial plant native to East Asia, which has traditionally been used as a tonic for physical strength and energy. The active components of ginseng are called ginsenosides, which are believed to be responsible for many of the health benefits associated with ginseng. Studies have suggested that ginseng can improve brain function, reduce fatigue and stress, boost immune system response and even help with erectile dysfunction.

Neem is a medicinal plant that has been used for centuries to treat various ailments such as skin disorders, gastrointestinal illnesses, fever, and malaria. Neem leaves contain compounds called limonoids, which are believed to possess antimicrobial properties, making them effective against bacteria, fungi, and viruses. Research suggests that neem extract may also be helpful in treating diabetes by reducing glucose levels in the blood.

Coca Leaves, native to South America, have long been used as an herbal remedy for numerous conditions. The active ingredient found in coca leaves is an alkaloid that has been found to reduce physical and mental fatigue, improve cognitive performance and enhance energy. Studies suggest that alkaloids can also help with depression, anxiety, and stress-related disorders.

Bitter Roots are a type of medicinal plant commonly used in African traditional medicine for their powerful healing properties. Bitter roots contain compounds such as quinine, gentianine, and iridoids, which are known to have anti-inflammatory, antifungal, and antioxidant effects. They may be especially helpful in treating malaria, fever, and gastrointestinal issues.

Fruits, vegetables, seeds, and nuts are packed with essential nutrients that provide numerous health benefits. Eating a diet rich in fruits, vegetables, seeds, and nuts can boost your immune system and reduce the risk of chronic diseases. Fruits, vegetables, and nuts are also rich in antioxidants, which help protect against inflammation, cancer, and heart disease. Seeds such as flaxseed are a good source of omega-3 fatty acids, which can lower blood cholesterol levels.

Herbal teas have long been used for their medicinal properties, with the herbs providing a range of benefits depending on what type you choose. For example, chamomile tea is calming and can help soothe digestive issues such as bloating or constipation; ginger tea may help relieve nausea; echinacea tea can boost the immune system; peppermint tea has antibacterial properties, and lavender tea is known to be calming and relaxing. The rooibos plant (pronounced ROY-boss) is made into an herbal tea filled with antioxidants. It is a South African flowering shrub used to make a mild-tasting reddish-colored tea that has no caffeine, very little tannin, and significant amounts of polyphenol antioxidants. This tea has been made in the Cedarberg Mountain region of South Africa for generations.

By including these natural foods in your daily diet, you can improve your health and well-being while providing protection against various illnesses. Natural remedies have been used for centuries to treat a variety of ailments and are still being studied today for their medicinal benefits. Whether you choose to eat them or use them in tinctures, teas, or salves, natural remedies provide a safe alternative to traditional medicine with fewer side effects. With the right combination of natural foods for healing, you can improve your overall health and well-being.

Finding the Right Natural Food

Finding the right natural food for your body's needs can be a challenge, as there are so many options available in today’s world. With the plethora of herbs, spices, and vegetables available, it is important to do your research before deciding which foods to include in your diet. It is also essential to consult with a healthcare practitioner or herbalist who can provide advice on how best to use natural foods for healing and what their potential benefits may be.

When considering which natural food will benefit you most, it is important to consider your health goals. Are you looking to improve your digestive health? Increase energy levels? Strengthen immunity? Knowing what you want to achieve with your diet will help you narrow down your choices and select the most suitable natural foods.

It is always a good idea to consult with a healthcare practitioner or herbalist when selecting natural foods for healing. They can advise on the best options available, considering any existing health conditions, lifestyle factors, and other medications you may be taking. A qualified herbalist can provide knowledge of herbs and spices that are not widely known in Western medicine.

Once you have chosen your natural food, it is important to research the potential benefits as well as any side effects they may cause. Different plants will interact differently with everyone's body, so it is essential to do your research and consult with an expert if needed. Common side effects of natural foods include digestive disturbances, allergic reactions, or changes in blood pressure. It is important to be aware of these potential issues before beginning a new diet regime.

By taking the time to consider your health goals and consulting a healthcare professional or herbalist, you can find the right natural food for your body's needs and begin to reap the benefits of its healing properties. With careful planning and preparation, you can make sure that the natural foods you choose are beneficial for both your physical and mental health. Practicing moderation and variety is key, as each nutrient provides its own unique benefits. With the right combination of natural foods for healing, you can create a balanced diet that will nourish and protect your body for years to come.

Chapter 8: The Need for Traditional Medicine

Traditional medicine, which encompasses a variety of natural remedies and therapies such as herbalism, acupuncture, and massage therapy, has been used for centuries to treat various ailments. With its holistic approach to healing the body and mind, it has become an integral part of many cultures around the world. In today's modern society, however, traditional medicine is becoming increasingly popular as people are looking for alternative solutions to their health-related issues.

This chapter will investigate why traditional medicine is so important in today's world and how it can bridge the gap between Eastern and Western healthcare systems. It will discuss the potential benefits of using natural remedies to treat illnesses instead of relying solely on pharmaceutical drugs. Additionally, it will explore how traditional medicine can be used to prevent future illnesses and promote overall well-being. Finally, this chapter will wrap up by examining the growing popularity of traditional medicine across the globe.

Importance of Traditional Medicine in Today's World

The world of traditional medicine is becoming increasingly relevant in today's modern society. It is an ancient practice rooted in various cultures around the world and has seen a resurgence in popularity over recent years. This form of treatment offers numerous benefits that are often overlooked by contemporary healthcare systems and makes it an invaluable tool to bridge the gap between Eastern and Western practices.

Traditional medicine has been used for centuries to treat ailments from simple everyday discomforts to serious illnesses. Currently, advancements in technology have opened new pathways that enable us to understand more about our bodies than ever before. However, these same advancements have also led to a certain degree of alienation from our more holistic roots. This is where traditional medicine can be invaluable in supplementing the treatments offered by modern healthcare systems. It offers an alternative view to healing and disease prevention.

Although Western practices focus on treating diseases after they have developed, traditional remedies are designed to be taken proactively to avoid illness altogether. It pays attention to the body's environment, both internal and external, which allows practitioners to identify potential risks that may trigger ill health or old age-related conditions. This approach can help to preserve overall wellness for longer periods of time.

In addition, traditional medicine is known for its natural healing properties, which can often provide long-term relief from a variety of conditions. Herbal remedies are renowned for their ability to improve circulation, reduce inflammation, and enhance immunity without the risk of side effects or dependency that may come with prescribed medications.

The prevalence of traditional medicine has been steadily growing in recent years as more people become aware of its potential benefits. There is an increasing demand for holistic treatments around the world, especially among younger generations who have grown up with modern technology and want to find alternative routes to better health. This has led to the development of many new practices and products that incorporate traditional remedies into their offerings, such as herbal teas, tinctures, and essential oils.

Benefits of Traditional Medicine

The holistic approach to healing with traditional medicine is one of its greatest benefits. Rather than solely focusing on the physical symptoms, this form of treatment considers the mental and emotional state of a person as well. It emphasizes that all aspects of a person must be considered to achieve optimum health. By combining various elements such as lifestyle changes, nutrition, relaxation techniques, natural remedies, and more, practitioners can create an individualized plan designed for each patient's specific needs.

Another advantage of relying on natural remedies instead of pharmaceutical drugs is that there is often no risk of side effects or dependency issues. Herbal remedies have been used for centuries to treat ailments from colds and flu to more serious conditions, and many of these treatments are still in use today. In addition, traditional medicine is typically cheaper than over-the-counter drugs and usually requires fewer visits to the doctor.

Finally, one of the most important benefits of using traditional medicine is its ability to help prevent future illness and promote overall well-being. This approach focuses on maintaining balance within the body rather than simply treating symptoms as they arise. It pays attention to environmental factors such as climate, diet, exercise, and stress levels that can have an impact on our health long term. By taking a proactive stance in preventing disease before it starts, we can reduce our risk of developing chronic illnesses or simply support our bodies in staying healthy longer.

Growing Popularity of Traditional Medicine

Plant-based healing is gaining popularity as people realize that many of the non-communicable diseases are a result of diets that are filled with processed carbohydrates and sugars. The ability to have some control over one's health has led to the new realm of food coaches; pre-boxed fresh ingredients for meal preparation delivered straight to one's doorstep have become all the rage. This, of course, follows evidence that foods rich in plants, vegetables and fruits are healthier options than many of the packaged foods found in modern-day supermarkets.

The growing demand for traditional medicine across the globe is evidence that this form of healing has become increasingly popular in recent years. This trend can be seen in both Eastern and Western societies, where more and more people are turning to natural remedies as a safer alternative to prescription drugs.

Recent research supports the efficacy of many traditional treatments, providing further proof that this type of care can be effective in treating ailments and improving overall health. In addition, there are now practitioners available all over the world who specialize in holistic healing with herbs and other natural remedies. This means that it's easier than ever before to find someone who can help diagnose a problem and provide an individualized approach to treatment.

The availability and increasing acceptance of traditional medicine are also made possible by the growing number of scientific studies that validate its effectiveness. Studies have found that certain herbs and other natural remedies can be just as effective as pharmaceutical drugs when it comes to treating some conditions. For example, research has shown that herbal teas can help reduce inflammation and improve digestion, while aromatherapy is often used to alleviate stress and anxiety.

The growing popularity of traditional medicine worldwide is further evidenced by many products now available for purchase. Herbal supplements, essential oils, tinctures, salves, and more are all accessible online or in health food stores. These items make it easier than ever before for people to access natural remedies without having to take a trip to the doctor.

Clearly, traditional medicine is becoming increasingly popular worldwide. With its ability to treat ailments without side effects or dependency issues, combined with its potential for preventing future illness and promoting overall wellness, it's easy to understand why people are turning to this form of healing more and more. Moreover, the availability of practitioners and products makes it easier than ever before for people to access natural remedies when they need them. Traditional medicine may just be the key to bridging the gap between Eastern and Western healthcare systems in today's world.

Chapter 9: Important Considerations

This book takes one on a journey across the globe, from east to west, sharing some of the historical perspectives and practices that are known as traditional medicine. Traditional medicine includes a wide array of practices, some of which are "cultural" to specific regions and groups.

The conclusion on traditional medicine based on evidence should be nuanced and consider various factors. Traditional medicine, which includes practices such as herbal medicine, acupuncture, and traditional healing methods, has been a part of human healthcare for centuries. Here are some key questions to consider:

Cultural Significance: Traditional medicine often plays a crucial role in many cultures and has been passed down through generations. In some cultures, certain roots and bark, for example, are boiled to alleviate menstrual cycle cramps (in northern Nigeria, the bark of the Neem tree, known as "*dogon yaro*" has been used for this purpose for generations. It is handed down from mother to daughter. It is deeply rooted in cultural practices and beliefs, providing a holistic approach to health.

Clinical Evidence: Many traditional remedies have proven to be effective in treating specific conditions. For example, some herbal medicines contain compounds with therapeutic properties. Unfortunately, there is no consistent evidence base for traditional medicines.

Integration with Conventional Medicine: In China, for example, traditional medicine is combined with conventional, western-style medicine to provide more comprehensive care to patients. The practitioners can harness the strengths of ancient wisdom as well as "modern" medicine. As a result, patient outcomes are maximized.

Potential for Harm: it is extremely important to note that many traditional remedies are generally safe when used properly (the right dosage, the right dilutions, the right methods, as applied by a practitioner); there can be harm if the previously listed items are not properly adhered to. Unfortunately, the lack of standardization for the preparation of many of the oral doses and limited laws and regulations of traditional practitioners can pose serious risks to people who may already be ill and in need of medical interventions. In some regions of West Africa, for example, traditional practitioners sometimes discourage patients from seeking conventional care for serious conditions needing urgent attention.

Regulation and Standardization: Although there is a global body known as the International Regulatory Cooperation for Herbal Medicines (IRCH), established in 2006, many traditional practitioners are in rural areas and unaware of such a body. [3] There are efforts underway in countries for the regulation of traditional practitioners and for the regulation of herbal medicines. To ensure the safety of patients, it is important that guidelines are established that will also control the quality of traditional medicines.

Patient Preferences: Many people in the southern hemisphere of the earth prefer traditional medicine. This is due to many reasons. First, an array of herbal medicines grows in the more tropical regions of the world. The environment has influenced cultural, traditional and spiritual practices. It is important for "modern-day" Scientists and Physicians to respect this and to understand patient – needs.

---

[3] World Health Organization. International Regulatory Cooperation for Herbal Medicine (IRCH). Retrieved from https://www.who.int/initiatives/international-regulatory-cooperation-for-herbal-medicines#:~:text=International%20Regulatory%20Cooperation%20for%20Herbal%20Medicines%20(IRCH)%20is%20a%20global,herbal%20medicines%2C%20established%20in%202006.

The role of traditional medicine in healthcare should be considered in a balanced manner. While some traditional practices have shown promise and cultural importance, it is crucial to approach them with a critical eye, relying on scientific evidence when available. An integrative approach that combines the strengths of traditional and conventional medicine while prioritizing patient safety is a potential avenue for the future of healthcare. Continued research, regulation, and collaboration between traditional and modern healthcare systems can contribute to a more comprehensive and effective approach to global health.

Conclusion

Traditional medicines and traditional practices often go hand in hand as rituals that bring together a healing practice with herbal medicines. They often cannot be separated and are part of a holistic approach to health.

It's important to note that the use of traditional medicine varies among different indigenous groups across the world. Many of the practices also depend on the natural resources that are available, encompassing flora, fauna, minerals and other objects used in ritual practices. In some communities, the sanctity of rain, streams and other bodies of water, the planets such as the sun and moon, are incorporated in these rituals and practices. These are often passed down through oral traditions.

There are challenges with climate change, environmental degradation and the use of natural resources that have led to questions surrounding the sustainability of certain plant species due to habitat destruction. This is leading to movements that promote ethical and responsible use of all the planet's natural resources.

There is a need for all the regions and cultures represented to be respected as knowledge that has been transferred from one generation to the next, in most instances through oral witnessing and practice. These cultures, traditions and the use of traditional medicines within those communities show wisdom and resilience in the face of physical, mental and spiritual health. The interconnectedness and a more holistic view are now apparent and evolving in "modern" medicine. It is essential that these cultures are preserved as they so often enhance current-day medicine. One of the aims of this book is to foster collaboration between healthcare practitioners and traditional medicine professionals, have an exchange of knowledge and enhance collaboration.

Traditional medicines have been used for centuries to treat physical, spiritual and emotional ailments for centuries all over the world. In recent years, there has been a renewed interest in this form of medicine as modern science begins to uncover the potential benefits of traditional medicines. This ebook examined the different uses of traditional medicines across the world and explored how they could be incorporated into modern-day healthcare.

The use of traditional medicines varies widely from culture to culture, but one thing is for certain—these remedies are essential for preserving human health and well-being. Traditional medicines offer natural solutions that can be used safely and economically by individuals who may not have access to or cannot afford conventional treatments. Furthermore, these remedies provide an opportunity to reconnect with ancient traditions while providing a sense of purpose and community to those who practice them.

It is important that we recognize the value of traditional medicines and work to preserve these practices for future generations. We must also be mindful of how these treatments are used and ensure that they are administered in a safe, effective manner. By doing so, we can ensure that everyone has access to the benefits of traditional medicines while minimizing any potential risks.

Additionally, it is essential to explore the potential use of these remedies in modern-day health care. While acknowledging their limitations, traditional medicines offer an alternative form of treatment that could help improve public health outcomes around the world. To maximize their effectiveness, research should focus on identifying active components and understanding more about their pharmacological mechanisms to enable evidence-based decisions.

Ultimately, traditional medicines offer a unique and invaluable resource for physical, mental, and spiritual healing. With the proper tools and knowledge, these treatments can be used safely and effectively to improve public health outcomes around the world. By preserving these ancient practices for future generations, we can ensure that every individual has access to the full benefits of traditional medicines across the globe.

To conclude, traditional medicines are powerful natural remedies with a long history of use in many countries around the world. These therapies offer an alternative form of treatment that could help improve public health outcomes while providing a sense of purpose and community to those who practice them. It is essential that we recognize their value and work to preserve these traditions for future generations. As we learn more about the pharmacological mechanisms behind these remedies, they could become an important part of modern-day healthcare systems. By doing so, we can ensure that everyone has access to the full benefits of traditional medicines across the globe.

Thank you for reading this book on Traditional Medicines with a glimpse across the World, from east to west.

I hope it has been enlightening and inspiring. With this knowledge, may you be empowered to explore how traditional medicines might enrich your own life and the lives of those around you.

Made in the USA
Monee, IL
07 November 2026

68012410R10039